Gerbils

Gerbils

WRITTEN AND ILLUSTRATED BY

Arnold Dobrin

Lothrop, Lee & Shepard Co. / New York

Printed in the United States of America
Library of Congress Catalog Card Number: 77-82101

Foreword

The science and practice of medicine is an ancient and honorable art in China. So when we asked for gerbils to be used for medical research, our request received prompt attention. Famers went out to the edge of the desert and trapped gerbils, for which they were paid the munificent sum of one-half cent each. A carpenter was called to build a box in which they could be shipped. The old man carefully contrived a beautiful little chest. He made it of white cedarwood, light and strong, because the shipment would be sent by plane. Each corner was cunningly joined without a single nail. He put two silver handles on the chest, each engraved with a goose in flight, to insure a safe journey. Inside he made a number of small compartments for the gerbils. This was the beginning of the journey that made the gerbil available for medical research.

While the gerbil has already demonstrated its value as an experimental animal in many fields of research, it was inevitable that it also achieve popularity as a pet. The gerbil is naturally so attractive and has so many interesting characteristics that it has become a favorite with many children in their homes and also in their classrooms, where it has been a subject for nature study, and in science classes.

This book, then, fills a real need for complete and accurate information on the characteristics and care of the gerbil.

Victor Schwentker
Tumblebrook Farm, Inc.
Brant Lake, N.Y. 12815
June 1, 1969

Meet the Gerbil

In 1954, Dr. Victor Schwentker brought twenty gerbils (pronounced JUR-bils) from Mongolia to the United States. At that time many people in this country had never heard of these animals. Since then gerbils have become popular as pets, and useful for classroom study and laboratory experiments.

There are many reasons for the gerbil's popularity. It is the small, appealing kind of animal that children are so fond of. It is also the kind of pet that some older people like. One elderly lady has reported that her young gerbil and old Irish setter are great friends. Sometimes the gerbil will crawl on the dog when he is sleeping, lift up one of his floppy ears, and go to sleep under it!

There are many species of gerbils. Mammologists—the scientists who study warm-blooded animals—have recorded that there are fifty-four different species in the world. Some of the popular names for them are "sand rats," "antelope rats," and "desert rats." The gerbil's scientific name is *Meriones Unguiculatus*, a combination of a Greek and a Latin word. *Unguiculatus* is a Latin word for fingernail. Some authorities believe that the gerbil's overly long nails are the reason for this part of the name.

All gerbils live in sandy, sparsely vegetated areas. They are found from southern Africa all the way north to Morocco and Egypt, in Israel and northern Arabia, and in southern Russia and Mongolia. Some species are sand-colored or mouse-gray. Others have bright reddish brown fur. But all have fairly long ears, elongated hind feet, somewhat round heads, and short, curved noses. Only the Mongolian gerbils are available in the United States.

The Mongolian gerbil is a rodent belonging to the family *Cricetidae* and the subfamily *Gerbillinae*. Some of its closest relatives are the hamster, the meadow mouse, the deer mouse, and the muskrat. Like other rodents it has incisors, front cutting teeth, which never stop growing. And, like other rodents, it is a mammal, which means that the mother feeds her young with milk from her own body.

Several features set the gerbil apart from most rodents. It has a short broad head, similar to that of a chipmunk, and when young, it has a little clump of fur on the end of its tail. These features not only make it different, but they also give it a very pleasing appearance.

A full-grown Mongolian gerbil is about four inches long. The tail adds another four inches. Adult gerbils weigh three to four ounces—about as much as thirty pennies. The fur on their backs is dark brown, changing to a very light brown on the sides. Their light gray undersides often look almost white. They have soft noses with whiskers of black and white.

Gerbils have short front legs and long hind legs that are strong and well-developed. Although their hind leg muscles allow them to take leaps of twenty-four inches, most of the time they run about on all fours. Each foot has five toes with relatively large black nails. The toes of the forepaws are so dexterous that the gerbil can pick up and handle small seeds. Consequently, when they eat, these animals often take a sitting position.

The Mongolian gerbil adapts easily to many different environments and gets along well in captivity. But this small animal is a very long way from home. Its native habitat is the Gobi Desert areas of northeastern China and Mongolia.

Survival in the Desert

In their desert homes gerbils must live through hot days and very cold nights. One of the most interesting things about this animal is the way it has been able to adjust to such extreme changes in the weather.

The gerbil has learned how to adapt to high temperatures with particular success. Experiments show that it can endure temperatures of 110 degrees Fahrenheit for as long as five hours. That is possible because the gerbil can regulate its body temperature in severe heat. In fact, this animal does it better than the kangaroo rat—a similar rodent, which lives in the southwestern desert of the United States.

The gerbil also avoids high or low temperatures by staying in the deep burrows it makes in the sandy soil. Here, in the darkness, temperatures tend to remain more constant than above ground.

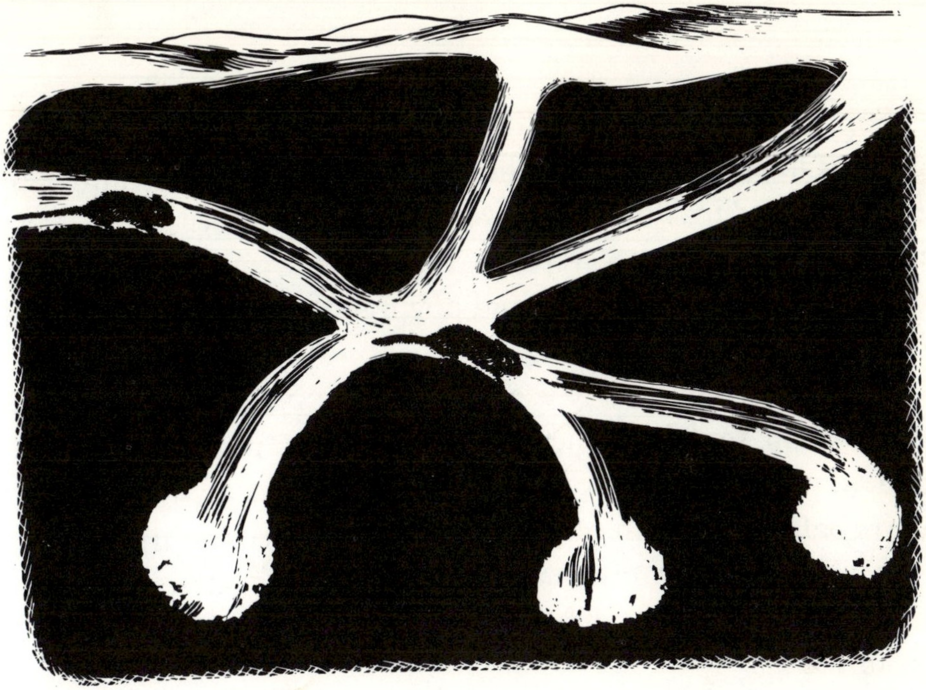

The burrows have many runways, which form a maze of connected underground tunnels. Some sections of the tunnels are used for the storage of food; others are for nesting. Within these burrows large colonies of gerbils live close together.

The sandy soil in which the gerbils construct their burrows also helps to protect them from their enemies. Shifting sands —moved by the wind and by the gerbils themselves—close the entrances of the burrows, making it hard for other animals to find them.

Since the Mongolian gerbil lives in wild, remote areas, no one knows for sure which animals are its natural enemies. But scientists believe that the gerbil is prey for snakes, hawks, eagles, foxes, and other animals that generally live on a partial diet of rodents.

In the wild, gerbils eat roots, grains, grasses, and seeds. Most desert areas contain juicy plants of various kinds and, from these, gerbils get the moisture they need.

Some writers believe that gerbils are active both day and night. Others report that they are active at night only, but tend to stay in their burrows on extremely cold or windy nights. One observer reports that gerbils are attracted to camp fires and will hop up and down in the outer edges of the firelight.

In the Metropolitan Museum of Art in New York City are three small sculptured jerboa mice, desert rodents that are related to gerbils. They were made almost two thousand years before the birth of Christ. A note describing the figures says: "These tiny creatures inhabit Egypt's desert fringe, where on moonlight nights they can be seen 'dancing' in groups."

The "dancing" may have been leaping. Both jerboa mice and gerbils have the long hind legs and shorter forelegs characteristic of many desert rodents. Much of the time a gerbil is content to scurry around on all fours, but now and then it will take a series of flying leaps. When it does, you will have a chance to see how well the long gerbil tail helps the animal keep its balance.

Gerbils as Pets

Gerbils are very active and curious animals. They are friendly and almost without fear. Although they may often sleep during the day, at night they are usually busy. They like to exercise on their wheels, shred material, rearrange their nests, or investigate any new object placed in the cage. Gerbils often alternate activity with short catnaps, but generally, they are always exploring, always investigating.

When you go close to the cage your gerbils will usually come up to you to discover what new thing or experience you may have to offer. If you stick your hand into their cage they will probably sniff it and explore it thoroughly. Often they will climb onto your hand, and if you are still, they will probably try to run up your arm.

The gerbil is famous for its curiosity. In one laboratory experiment hungry gerbils were put into a maze of paths leading to food. But instead of trying to find the most direct path to the food, the gerbils preferred to explore several different paths along the way.

Any strange object you put into the cage will be chewed or gnawed. Gerbils seem to be constantly searching for something to gnaw on. They can easily eat through many types of plastics, some woods, and even tin. Never let them loose in your closet or you will soon find holes in all of your clothes!

Gerbils may spend a large part of the time gnawing on the bars of their cages. At first, this may look like a desperate attempt to win freedom. They are simply trying to keep their teeth sharp and prevent them from growing too long. If you see them chewing on the bars of their cages a good deal, it may be wise to give them a block of wood or a small branch from a tree to gnaw on. Gnawing on sharp grains of food also helps to keep their teeth from growing too long. This is the way that gerbils, like all rodents, keep the sharp teeth they need for survival.

Most of the time these animals are silent but once in a while they will make squeaky little noises. The only other sound they make is with their feet. This is a high, drumlike beat made by rapidly stamping one hind foot against the floor. This sudden TAT-TAT-TAT may be a warning signal. In any case, it seems to indicate excitement. Most researchers believe it may have something to do with fear, anger, or sexual activity.

Gerbils keep their burrowing instinct in captivity and will often shape large pieces of fabric into the form of tunnels. A good part of their time is then spent scurrying around these burrowlike shapes. It is the gerbil's forefeet which do most of the digging; the hind legs push the material they dig to the rear.

These small animals are very sociable. They do not seem to like living alone, and are happiest in pairs or groups.

Usually the males and females live together peacefully. Most of the time two females will live together peacefully too. When fights do occur, though, it is usually between adult females. Generally two males will live together without problems. If they do start to fight, they, like females, must be separated at once.

One pair of males that was put into a cage together cautiously stayed on opposite sides of the cage from each other for two days. They observed one another, but scampered apart when they found themselves getting close. Finally they seemed to make up their minds that they liked each other. After that, they began to sleep contentedly side by side in their nest.

How to Handle Gerbils

Gerbils have no fear of human beings. Since they are gentle animals even in their wild state, there is no question of "taming." One of the reasons they make such good pets is that they are so easy to handle. Often it seems as if they actually want to be picked up, and they like to take food from your hands.

They are also very quick. Trustful but curious, they will run along your arm and search your pockets if they suspect food is there. If it is, they will continue to do this. Sometimes this is called a "trick." But the gerbil is just on its continual search for food, just as it would be in an underground maze or along paths through vegetation in the wild.

Scientists pick up gerbils in laboratory work by getting a firm grip on the tail as close to the body as possible. Be very careful when picking your gerbil up this way.

Another way is to place your hands near the gerbil in a cupped position, palms up. Let it climb into your hands by itself. Then you can stroke its fur and let it get used to you.

Always be gentle. Gerbils will bite if they are mishandled —especially if surprised by a sudden squeeze. Any scratch, whether it penetrates the skin or not, should be washed with

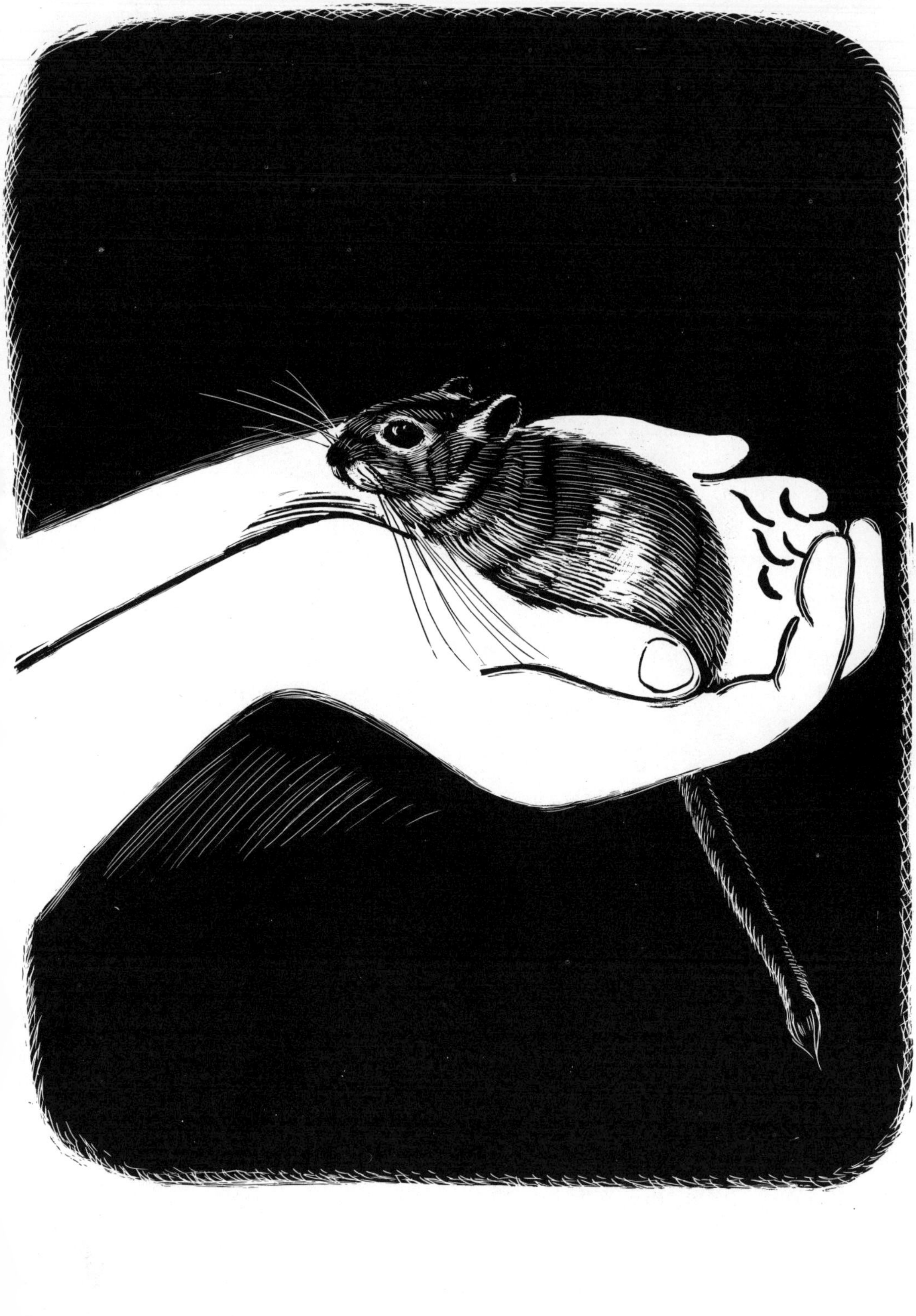

soap and water and treated with an antiseptic. Once in a great while you will find a gerbil that bites frequently.

Keep your gerbils away from high ledges and table tops. Scientists say that gerbils do not seem to have a good sense of heights and will stray too far to the edge and fall off.

It is best to begin handling your gerbils when they are about six weeks old. (And never pick up a gerbil pup until it is at least one week old.) Move slowly and calmly when you handle them. Do not grasp them too quickly or hold them too tightly. When they eat from your hands, gently stroke or scratch their backs, heads, and ears with your fingers. Then the next time they will be sure to approach you with less fear.

Fun With Gerbils

It is great fun to take your gerbils out of the cage. Let them explore the room and scurry around. They will quickly scamper through cardboard boxes, oatmeal containers, and cardboard tubes, pushing them around and nibbling them. You may be surprised to see how willingly they come back to the cage. Gerbil experts say that gerbils do not seem so eager to escape as other rodents. A sure way to lure them back to the cage is by spreading sunflower seeds, like stepping stones, up to the door. Or spread the sunflower seeds outside and inside a wastebasket turned on its side. As soon as the gerbils enter the basket, you can scoop them up and return them to the cage.

An amusing thing to do is to put your gerbils in a dollhouse and watch them run from room to room, investigating. You might hide some sunflower seeds around the dollhouse and under the furniture. Remember, though, that both dollhouse and furniture can be destroyed if the gerbils remain there for any length of time.

After you play with your gerbils, always be sure to return them to their cage. The United States Fish and Wildlife Service and the Department of Agriculture are worried about what might happen if gerbils were allowed to breed in the wild. For this reason, some of the western states have laws forbidding the possession of gerbils.

Such laws are not aimed at the people who wish to keep a few gerbils but are preventive measures intended to keep gerbils from fields and farms, where they might cause serious damage.

It is probable that foxes, owls, hawks, and the other animals that feed on rodents would keep the gerbil population down. Nevertheless, history has shown that the introduction of an animal into a new environment may result in unexpected difficulties.

When English sparrows were brought to this country over a hundred years ago, it was hoped they would destroy harmful insects. Instead they create problems for our native birds.

To prevent escaped gerbils from becoming a problem, take care to see that they remain in their cages except when you are playing with them. Although it is hard to think of the fascinating gerbil becoming a pest, precaution is still a wise policy.

Choosing Gerbils

When you first see a group of gerbils they will probably all look very much alike. Healthy gerbils have thick, soft, glossy fur, firm, well-fleshed bodies, and sparkling bright eyes. They move quickly and scurry around the cage when some object is placed inside. It is only later, after you have had a chance to observe your animals, that you will be able to detect differences in gerbil behavior. The important thing to remember when buying or choosing gerbils is that quiet or sluggish-looking animals may be ill.

Those with sores—often the result of fights—should not be chosen. Excessive scratching may mean that the gerbils are carrying parasites, such as fleas. Bald spots indicate that they have not been properly fed.

Gerbils can be bought in most pet shops. If the shop you usually go to does not have gerbils, the shop owner will probably know of another that does. These rodents were difficult to find a few years ago. They have become so popular, however, that many pet shops now stock them.

Often people who have been keeping gerbils for a while have more animals than they want. Ask your friends and teachers if they know of someone who has gerbils for sale. If purchased in this way, they may be cheaper than they would be if you bought them through a store. Prices vary in different parts of the country but they usually cost between $2 and $5.

Gerbils that are six or eight weeks old are probably the best age to buy. And since these animals prefer to live in pairs, it is best to buy two. If you do not wish to breed gerbils, buy either a pair of males or a pair of females.

Mealtimes for Gerbils

Although gerbils are mainly herbivorous, they often eat other foods that are offered. They will eat insects or worms, although they need no animal flesh of any kind to remain healthy. That they can adapt themselves to a diet made up mostly of animal flesh was proved by a Japanese scientist who daily fed his gerbils small dried fish (in addition to grains). Gerbils will try almost any new food that is offered to them, but their favorites are grains and seeds.

Captive gerbils will even thrive on a regular diet of birdseed. The easiest way to feed them, though, is with packaged foods that are prepared for hamsters, mice, and other rodents. These are balanced to provide the best nutrients a gerbil needs—a mixture of corn, wheat, barley, and various seeds. Most of them also contain a meat or fish ingredient. These prepared foods are found in all pet shops and most markets and ten-cent stores.

Gerbils can be fed at any time of the day. About one tablespoonful of food once a day ought to be enough for the average adult gerbil. Half of that amount is enough for a younger gerbil. Food can be scattered on the floor of the cage

or put in a special container. This may be an ordinary dish, saucer, or one of the wire mesh or slotted metal food hoppers sold in pet shops. However, these are not necessary. A little pile of food on the floor of the cage tastes just as good to a gerbil.

Captive gerbils are as curious about new foods as they are about new things and places. Part of the fun of keeping gerbils is seeing their reaction to new foods. They will eagerly eat such processed foods as crackers, potato chips, cereals, and popcorn. Dog biscuits or dry cat pellets may also be used. Vegetables and fruits that gerbils especially like are celery, kale, carrots, potatoes, and apples. Gerbils will also eat green leafy vegetables but those which are more crisp are better for them. Too much soft food is not good. Gerbils in the wild

state naturally keep their teeth from growing too long by eating hard grains.

Milk and bread soaked in milk will vary your gerbil's diet. One way to help keep young litters healthy is by giving a mother condensed milk when she is nursing.

When You Go Away

Gerbils will not hoard food and, except with sunflower seeds, they will not overeat. Because of this you can give them enough food for several days without fear that they will stuff themselves. This makes the gerbil a perfect pet for families that travel a lot.

You can fill their containers with enough food and water for a week and be sure they will be getting enough to eat at the end of that time. But it is a good idea not to leave them alone any longer than a week. Food can spoil, vegetables dry up, and sometimes a water bottle will leak.

Like many other desert animals, gerbils can go for long periods of time without water. Captive gerbils—wild ones—can survive entirely without water if enough fresh, juicy vegetables are included in their diet.

Some Indian cousins of the Mongolian gerbil were given a completely dry diet by Indian scientists who wanted to find out why they could survive so long without moisture. The experimental gerbils were given no water or juicy foods of any kind. The biologists discovered that their gerbils could go as long as four months on a completely dry diet.

The main reason for their amazing endurance is their strong kidneys, which are able to remove salts and wastes in a very highly concentrated urine. This allows the gerbil to keep most of his body moisture and release only a small part of it every day. As the biologists found, the gerbil's blood becomes thicker when it gets little moisture, but continues to do its job.

The best way to provide water is in an inverted water bottle with a rubber stopper and a curved metal tube. Any pet supply shop sells them. With this bottle securely attached to the outside of the cage, your gerbils can have a clean sip of water whenever they want it.

Such a bottle cannot be tipped over as a saucer can. With a saucer, there is always the danger that a baby gerbil will fall in and drown. Also, a gerbil will usually fill the dish with wood chips or whatever bedding material is in the cage. The large damp mass that results is neither very pleasant to look at nor very healthy.

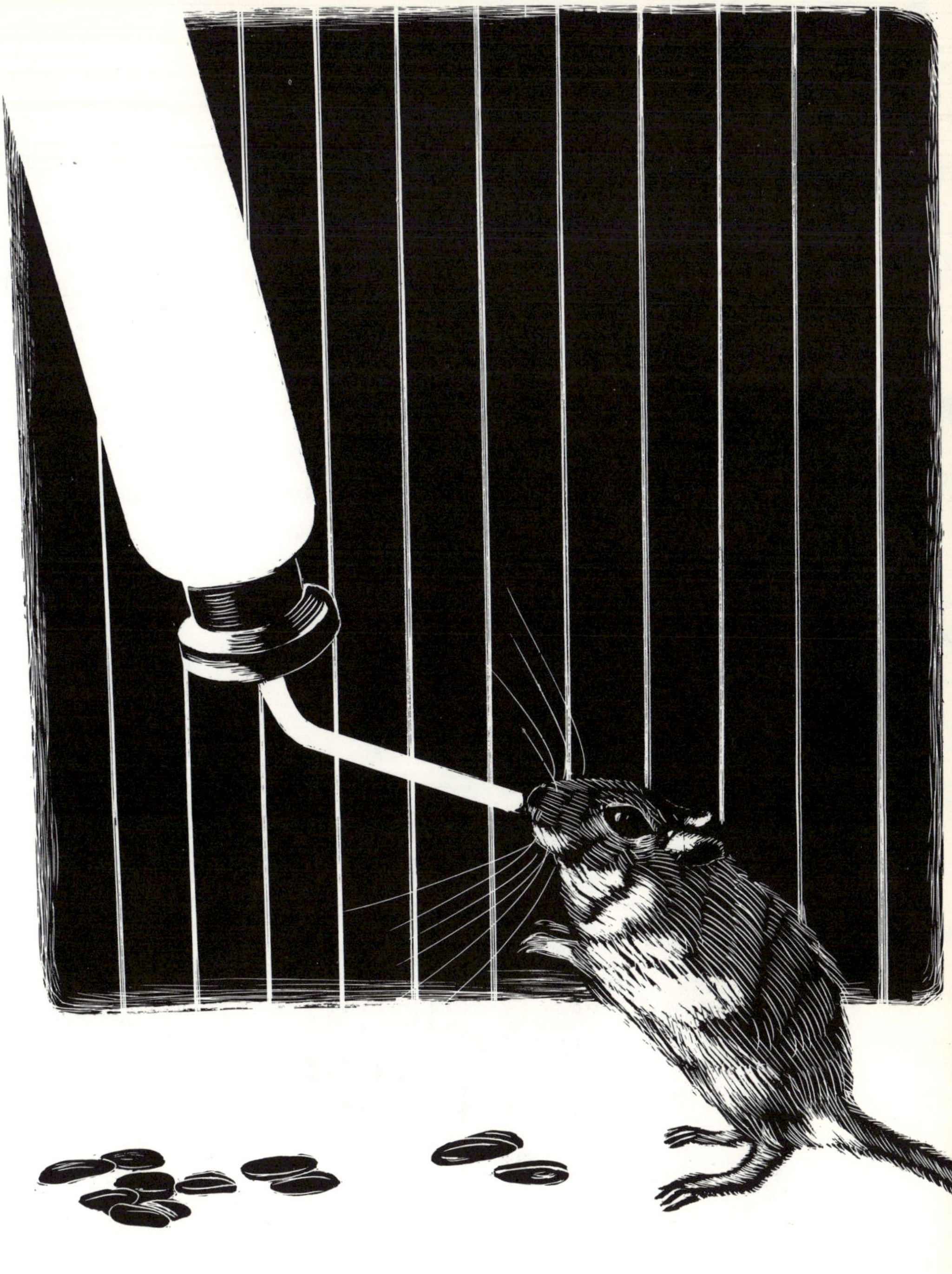

Homes for Gerbils

Simplest of all is the standard small metal animal cage which comes in two sizes. Both sizes have a shelf, an exercise wheel, and a movable floor. These cages are also equipped with an inverted water bottle.

Their main disadvantage is that busy, active gerbils are sure to scatter some of their bedding material through the bars. This, of course, makes a mess around the cage.

To prevent this, get a corrugated cardboard carton from the grocery store. Cut it down to about four or five inches from the bottom and place the cage inside it. The carton need only be slightly larger than the cage. It will be more attractive if you paint the outside of it a bright color.

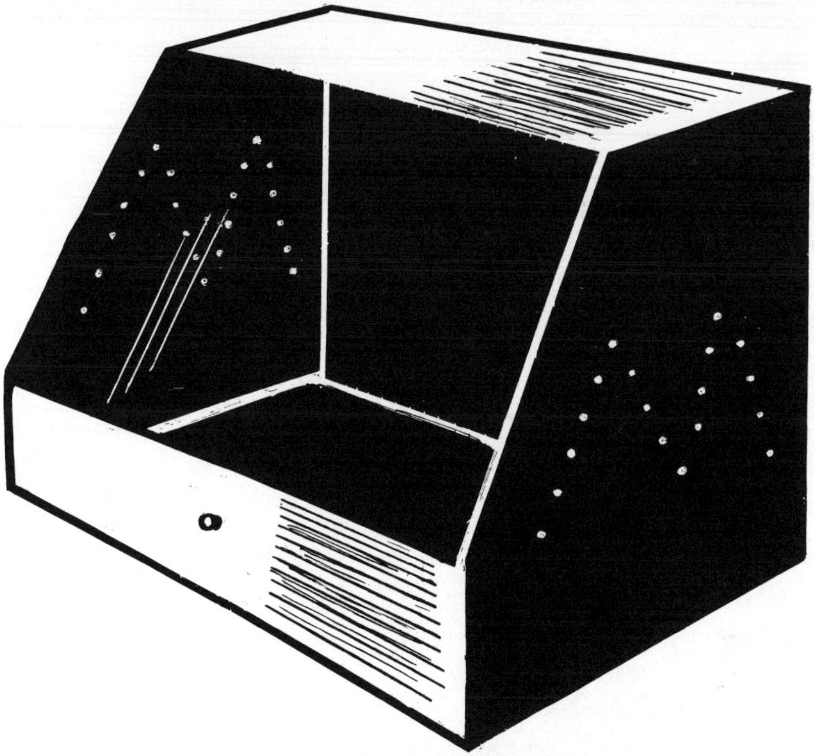

Cages of clear heavy plastic also solve the bar problem but they are usually more costly than metal cages. Although gerbils can chew through most plastic, they are not able to get a hold on the smooth walls of these plastic cages.

An aquarium also makes a good cage for gerbils. Cracked or slightly damaged aquariums can sometimes be bought at pet shops very cheaply.

If this type of cage is used, put a secure wire mesh across the top—especially if there are other pets in the house. This should be one-half-inch (or less) wire mesh; it is sold in most hardware stores. Avoid insect screening, which can be easily chewed by sharp gerbil teeth. Get a piece of wire mesh screening large enough to allow it to extend about two inches beyond each side of the top of your cage. Then just bend it over. If an aquarium is used, it is a good idea to buy an exercise wheel for it.

If you decide to house your gerbils in a homemade wooden cage, use a very hard, thick piece of wood. Otherwise sharp gerbil teeth will soon be gnawing through the walls.

A homemade cage can have wooden walls on three sides, with a wire mesh securely nailed to it to make the fourth side. Fit another sheet of wire mesh over the top, one that can be removed. Then you can take the gerbils out easily.

The size of the cage should be 10 x 18 x 10 inches, or 12 x 18 x 10 inches.

Because gerbils often sit up on their hind legs when they eat, the walls *must* be at least 6 inches high. Walls that are 10 or 12 inches high are better, though. They allow room for leaping.

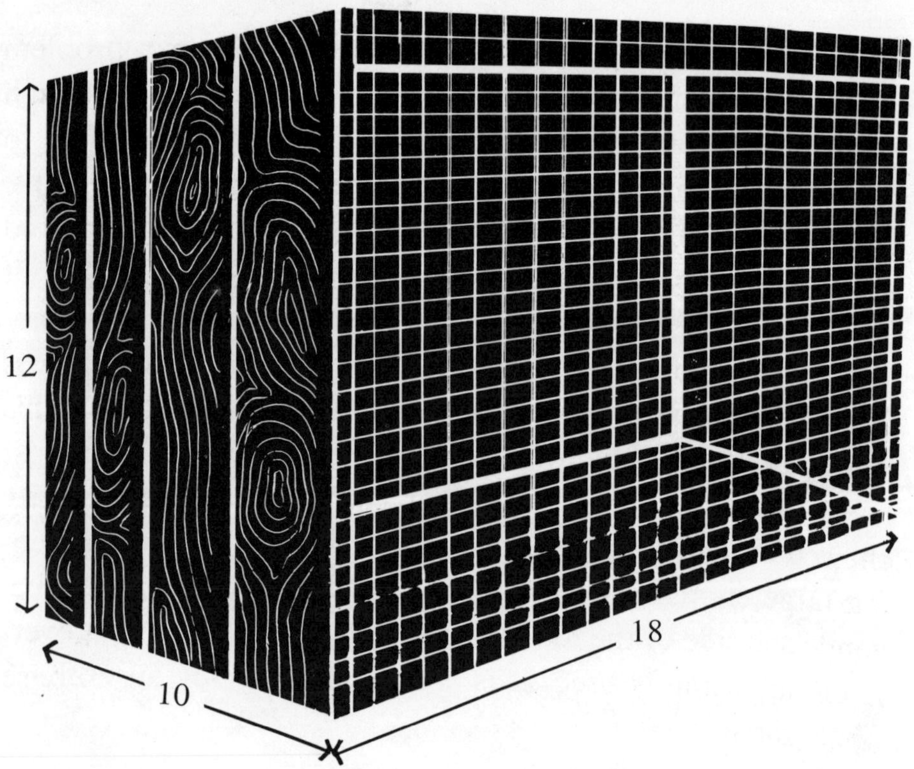

The easiest way to provide water in a homemade wooden cage is by first drilling a small hole through one side of it. The water bottle can then be easily attached to the outside while the stem is fitted into the hole. Using this method, you will always know when your gerbils need more water and can refill the bottle with little difficulty.

The floor can be of wood, but you should be able to remove it, if possible. This will make it easier to clean the cage. In a few minutes you can slide out the floor, wash it off, and slide it back in again.

If you want a nice roomy cage, you can make one two-stories high. Make a hole in the top floor, and fasten a ladder to it. Then your gerbils can run upstairs and downstairs. Make it the same way with the same dimensions except for height. Nail the second floor in securely, about halfway up.

Never place gerbils in direct sunlight. Continued exposure to strong, hot sunlight can kill them. Place the cage in a position that will give the animals both light and shade. The ideal temperature is 65 or 75 degrees Fahrenheit, about the same temperature that most human beings find comfortable.

Gerbil Beds

Gerbils will use just about anything for bedding or nests—as long as they can chew it, push it around, and make it snug and warm. If there is a shelf in the cage, gerbils often make their nests under it. When you clean the cage, discard the nests and they will promptly make fresh ones.

Pet shops sell many different kinds of packaged bedding materials. However, straw, various papers, dry grasses, old clothes, ground corncobs, or any kind of wood shavings are just as good. Often gerbils will use pieces of cloth to line their nests.

(One person, watching some gerbils punch holes down the length of a piece of paper and then pull it into strips, said they

work like bus or streetcar conductors punching transfers. Gerbils often work with paper this way.)

Dirt or dirt mixed with sand can also be used for this purpose, but this makes the cage harder to clean. One of the interesting things about using sand and dirt is that you can see the way gerbils actually burrow through it as they might in the desert.

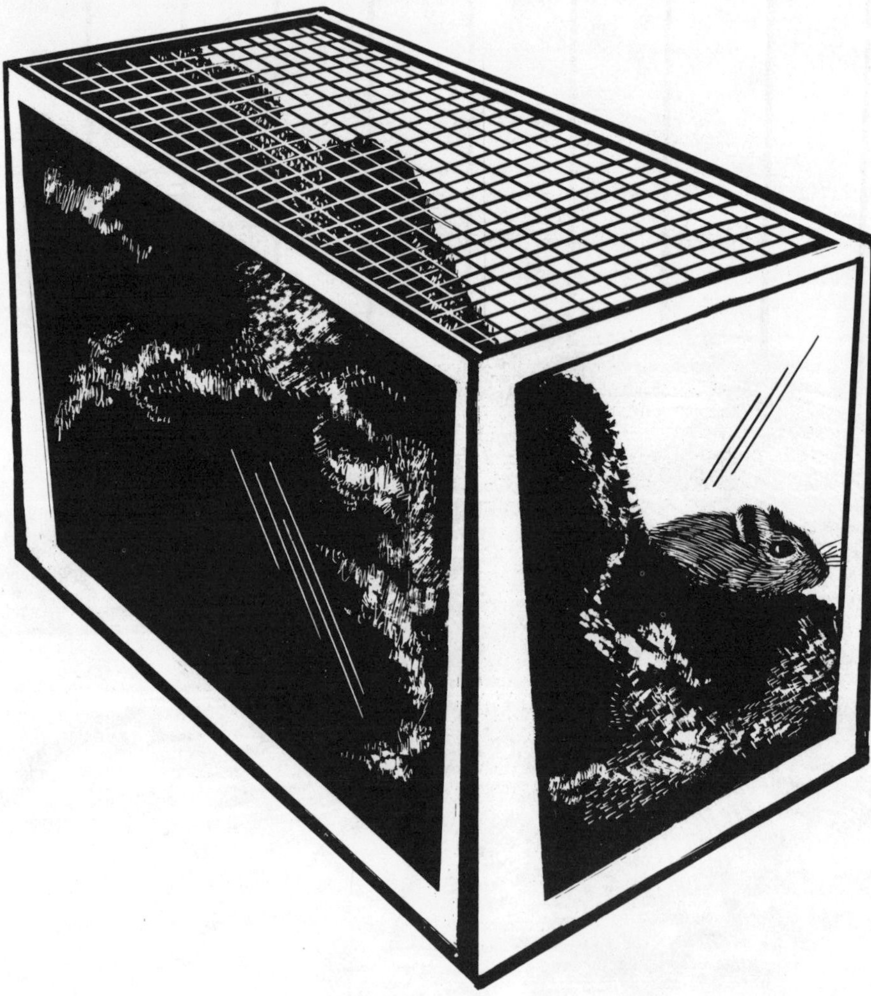

Keeping the Cage Clean

You will notice that gerbils often set aside a special place for their excrement. An interesting experiment is to drop a handful of seeds or crackers in this area. One by one, the gerbils will carry them to another corner of the cage and they will eat them there.

Because gerbils drink such a tiny amount of water they excrete only two or three drops of urine a day. Their droppings are also very small and almost entirely without odor. Because of this, and because gerbil cages remain so dry, they need to be cleaned only twice a month.

In warm weather the cage can quickly be hosed down in the yard, then wiped dry with a newspaper.

Indoors a vigorous rubbing with dampened newspaper, paper towels, or rags will do a good job. Three or four times a year the cage should be cleaned with a household disinfectant. However the cage is cleaned, remember that it should be perfectly dry before you put fresh bedding material in it.

Gerbil Families

It doesn't take long to find out which gerbil is a male and which is a female. The easiest way is simply to pick them up carefully by their tails, as close to the body as possible, and examine the area underneath.

The genital opening of the female is close to the anal opening—about 1/8 of an inch away. In the male the distance is about 1/2 inch. The coloring of the male's scrotum—the pouch

male

which holds the reproductive organs (testicles) containing the male sperm—is darker than the fur around it. Usually this makes a slightly bulging appearance above the animal's hind legs, under the tail.

Males are often slightly larger than females and sometimes more aggressive. Often the male will take what food he wants first, and he may push the female away from the water bottle when he wishes to drink.

Gerbils do not mate as often as rats or mice. Still, a female may produce ten litters of young in an average lifetime—about four years. (One gerbil is known to have lived five and a half years.) It is believed that wild gerbils breed throughout the year but produce the greatest number of litters in the spring and autumn seasons. Captive gerbils will breed at any time.

female

For best results in breeding, make sure that the gerbil parents have a roomy cage and enough nesting material. Place the cage in a fairly private place, away from too much noise and disturbance. Give the pair a well-balanced diet and see that the female has milk every day. After pups are born, do not handle them, especially during the first week after birth.

Unlike hamsters, the male and female may be left together when the female is carrying her young and afterward as well. In captivity gerbils are monogamous—meaning that they will accept only one mate. (Researchers are not yet sure if this same pattern is also followed in their natural habitat.) Occasionally a gerbil that has lost its mate will take another but this seems to be the exception. Female gerbils will sometimes completely refuse to accept another mate.

It takes about twenty-five days for the baby gerbils to grow inside the mother. However, it is hard to know when a gerbil is pregnant. Probably the first sign of the new gerbils will be their tiny squeaking noises. Most litters are born at night or in the early morning. There can be only one in a litter or as many as twelve. Usually there are four to six.

Even while a gerbil mother is nursing it is possible for her to have another litter. If this happens, she has two sets of babies to feed. Some of them will probably die because it is unlikely that she will have enough milk for all the pups. So keep a nursing mother separated from the male.

The gerbil mother will not make a special nest for her babies. After they are born, though, she will spend most of her time crouching close to the pups to make sure they are warm. Most mothers watch their young very closely and see to it that none of them stray far from the nest.

Some researchers claim that the male will help in caring for the young. Others report that he does not help, but that he will not hurt them, as hamster males often do.

The mother will usually allow you to pick up her pups without biting or protesting, but this is still not a good idea because it puts the young ones in danger from their own mother.

She may—like the female rat—attempt to eat her young when they have been disturbed. This is especially true of the first week when the pups should be left entirely in her care. Sometimes a female will kill all of her young without any reason that we can see. Sometimes an entire litter may die simply because she is a bad mother—but it often happens only to the first litter. After that, the mother seems to know more about raising a family and their chances for survival are better.

Gerbil Babies

When the babies are born they have no hair and are a pale pink color. They are also blind, toothless, and deaf. Their toes have soft, colorless claws. But gerbils grow fast and by the end of the first week they are crawling around the cage.

A fine fuzz soon begins to cover their small pink bodies. When they are two weeks old the fur has already thickened and they begin to look like small copies of their parents. By the time their eyes are open the fuzz has changed to a beautiful coat of soft, light-brown fur. When the babies are three weeks old they will be able to see.

They are already curious. They begin to explore the cage now, crawling and running into all the corners. They tumble and wrestle with their brothers and sisters and begin to eat solid food.

Usually they will nurse as long as their mothers permit them. They may still seem young and helpless but they should be weaned when they are three weeks old. If you allow the mother to go on nursing, she will probably keep it up until they are about six weeks old. To wean the pups, just put them in a separate cage and start giving them the same food you have been feeding their parents. (Until they reach maturity they can remain in a smaller cage.) This is also the time when you will be able to find out what sex the pups are.

Once in a while a litter will have one or two runts—unusually small animals. Don't be surprised if they die within several days after their birth. If they survive this period, the chances are that they will catch up to the other animals in the litter and become healthy adults.

Keeping Gerbils Healthy

Gerbils are healthy animals. Most untimely deaths are due to fighting, and this can be avoided if you are watchful. Generally, you will find out very soon if your gerbils do not get along with each other. Those that start to fight should be separated. If they get along well, you can be pretty sure there will be no bad fights even if you leave them alone as long as a week.

These animals carry no diseases of their own. And since they will probably have no contact with wild animals, it is very unlikely that your pet gerbils will catch a disease such as rabies.

Sometimes a gerbil will have a runny nose or eyes and lose its appetite. This is probably caused by a cold; it will pass in time just like an ordinary cold in a human being.

Sometimes a gerbil will develop a bald patch. If you make sure that his diet is well-balanced, this will almost always disappear in a week or two.

Gerbils can carry fleas and lice. If they are doing a great deal of scratching or are biting their fur, use a dusting powder or spray to delouse them.

Any body sore or eye injury from contact with sharp objects or from fighting can be treated with an ordinary antiseptic. Eye injuries will respond quickly to an antibiotic eye ointment or to rinsing with a disinfectant solution. The ointment must be quickly and carefully squeezed from the tube into the gerbil's eye; the solution can be dropped from an eye dropper.

Gerbils sometimes have "fits" or "seizures" similar to the seizures of epileptic human beings. It is believed that these fits may be due to strange cages, sudden noise, or rough handling. Probably there are other reasons which scientists have not yet discovered.

Researchers in the Department of Psychology at the University of Texas found that 50 per cent of their gerbils suf-

fered occasionally from an epileptic type of seizure when placed in a strange and unusual environment. However, this tendency of gerbils should not keep anyone from having them as pets. Such "fits" pass quickly and in no way affect the gerbil's health, which is usually excellent.

Gerbils are clean animals. Often they use their tongues and forepaws to clean their faces, heads, and bodies. Sometimes you will see them cleaning each other's bodies. There is no need for gerbil owners to ever clean their animals—only their cages.

Gerbils in the Research Laboratory

Today scientific researchers use a large number of different animals in their laboratory experiments. For many years the rat, rabbit, and rhesus monkey have proved their worth as valuable experimental animals. Although these animals are still in great demand, scientists are constantly looking for new animals to use in research work.

When Dr. Schwentker brought the first gerbils to this country he wanted to test them as laboratory animals. At that time he had no idea that they would become popular as pets. After a few years of experimenting with these animals in the laboratory he called them "an essential tool of the research worker."

Today a wide variety of scientists in many parts of the world use gerbils in research. These are some of the fields in which gerbils have proven useful: endocrinology, the study of the endocrine glands which produce vital hormones; hematology, the study of the blood and blood-forming organs—the spleen, and bone-marrow; and parasitology, the study of parasites.

Perhaps the field in which the gerbil has become most useful is psychology. Gerbils learn very quickly. They learn what psychologists call "avoidance response" about ten times as quickly as rats.

In one experiment they were put into a box which had a fence across the middle of it. When a button was pushed, an electrical impulse was sent into one half of the floor. As soon as the gerbils felt it, they jumped over the fence to the other side. To avoid the electrical charge, they kept jumping back and forth. When both sides of the box floor were charged with electricity at the same time, the gerbils jumped to the top of the fence and stayed there.

A gerbil's eyesight is keener than that of the mouse or the rat. They can hear sounds of low frequency, which cannot be heard by the human ear. Unlike white rats, which stay close to walls and fear being out in the open, gerbils seem to be unafraid. All of these characteristics have aroused the interest of the psychologists.

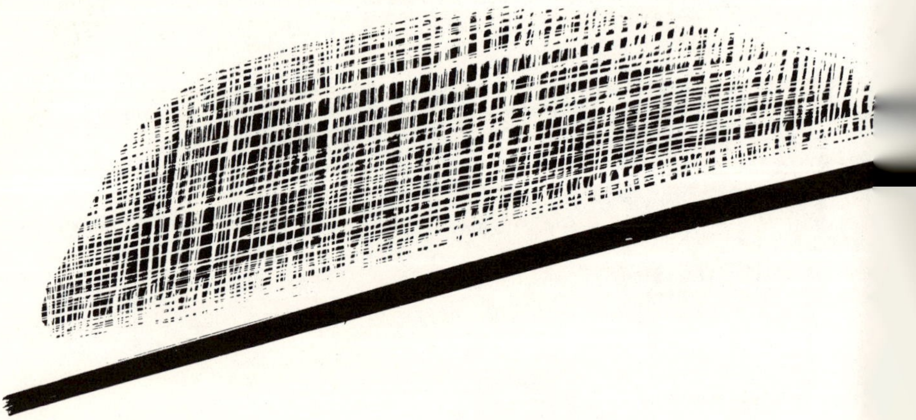

Observing Gerbils

It will help to study your gerbils if you can tell them apart, but this can be difficult when a group of babies grows into adults. The problem is solved by just placing a drop of food coloring—or a slight smear from a felt-tipped pen—on their foreheads. These colors will not hurt them and will make it easier to identify them quickly.

One good way to start studying your gerbil is to compare it to other small rodents. Keep a notebook and write down your discoveries and observations. After a while you will find that your notes have helped you understand your pet much better.

If you are lucky enough to have hamsters, mice, or other rodents to observe, put their cages next to the gerbil cage. Compare the gerbils with the other rodents. Notice the differences of tails, legs, heads, and the way they move about in the cage. Find out which rodents are quick, which sluggish, and which alert.

Notice whether your gerbils spend a lot of time in a sitting position or sit only when they eat. Observe how they rest. Do they lie on their sides or sleep on all fours? Watch them carefully when they are drinking, eating, and grooming each other, and write down everything that you would especially like to remember.

It is fun to give your gerbils different kinds of things to climb on. Try some things that are easy and some that are hard—such as slippery plastic tubes. Watch the gerbils attempt to climb each of these objects and see whether they succeed.

Paw Prints

To begin this interesting experiment you must first make a wide flat bed of wet smooth sand. This can easily be done in a cut-down corrugated carton. It is not permanent but it is not intended to be.

Put your gerbils down on the sand in the carton and look at the delicate paw prints they make. Compare these to the paw prints of other animals. Then carefully examine the gerbil's paw. Notice the special features it has for moving in dry sandy soil.

Changing Your Gerbil's Environment

This experiment will show you how the gerbil lives in different kinds of environments. You will not find that there are any big differences, but you will be able to see how gerbils make use of the different material they are given.

If you have several cages available, they can all be used at the same time for this experiment. Otherwise, the same cage can be used but you will have to change the materials in the cage every week.

To begin with, you might use only pebbles, rocks, and sand. Watch how your gerbils move the pebbles—how they work in and around the rocks. Do they climb the rocks and sit on them? Of course they will burrow in the sand. Let your gerbils make their home with these materials for a week and then clean the cage.

Next you might use piles of dry leaves, grass, and small pieces of bushes and flowers, perhaps cacti or other succulent plants from their native habitat. These will probably be eaten by the gerbils, but of course that is part of the experiment.

Another environment might be made of only rags and paper. A cage with a good supply of rags and paper is sure to keep your gerbils busy chewing and shredding. Notice the kind of nests they make out of these man-made objects. See whether they look very different from the nests made out of leaves and grass. Notice, too, whether the gerbils carry their nesting material from one place to another and whether the females and males work together to make the nest.

Watch how your gerbils react to each of these environments and mark down your observations in your notebook.

Choosing Their Own Food

You can discover what foods your gerbil likes best by using what is called "the process of elimination." With this method you will discover that your gerbils do not like some foods nearly as much as others. Eliminate, one by one, the foods that they choose to eat *last*.

Start by offering two different kinds of food at the same time, placing them about five inches apart. They can be any food you choose. Make sure that you always use the same amount of each food and that each is always offered in the same place. Watch which food your gerbils eat first. On the next day give him more of the food that he likes *plus a new food*.

On the following day repeat what you did before. Give your gerbil a pile of the food he ate most quickly with another pile of entirely new food. Write his preferences in your notebook. Perhaps you will discover one or two foods (in addition to sunflower seeds) that your gerbil likes more than any others.

You can experiment with all of the dry cereals, seeds, grains, solid and leafy vegetables, nuts, fruits, bark, roots, leaves, and all of the packaged foods for hamsters, rats, guinea pigs, dogs, cats, mice, or birds.

Other foods that are interesting to try are meal worms, meat, eggs, fish, and insects. Most gerbils will at least nibble them. Some will eagerly eat them, while some may refuse them entirely.

How Sharp Are Their Teeth?

It is interesting to test your gerbil's ability to gnaw through different materials. Try giving him a different thing to chew on each week. You might start with small branches or twigs of different sizes and shapes. Then try plastic toys or containers and pieces of heavy cardboard or leather. Examine your gerbil's teeth each week to see how these different materials affect them.

Gerbil Explorers

If you want to study the gerbil's curiosity and need for exploration, put the gerbil cage in the center of a room in which all of the doors have been closed to avoid losing it. Do not open the cage just yet. First, set up your room so that the gerbil will have a few things to explore.

At different distances from the cage place several small boxes or cans. Those that are big enough to allow a gerbil to crawl inside can be left in an upright position. The others should be put on their sides. In some of these containers put different kinds of foods.

Then open the door of the gerbil cage and stand quietly to one side of the room and watch. Do they appear to be fearful at first? Do your gerbils stay close to their cage in the beginning and only gradually begin to explore, or do they bravely start out at once to see the world outside their cage? Notice whether they go directly to the closest containers or explore those farthest from the cage first.

You will probably find that you like some of your gerbils more than others. You will learn that even though at first they may seem to be exactly alike in their behavior, they can really be very different. But whether you study them or not you will certainly discover that the gerbil is a delightful little pet and a fascinating animal.

Index

1 2 3 4 5 73 72 71 70